Albert Salisbury

Phonology and Orthoëpy

an elementary treatise on pronunciation for the use of teachers and

schools

Albert Salisbury

Phonology and Orthoëpy
an elementary treatise on pronunciation for the use of teachers and schools

ISBN/EAN: 9783337406417

Printed in Europe, USA, Canada, Australia, Japan

Cover: Foto ©berggeist007 / pixelio.de

More available books at **www.hansebooks.com**

PHONOLOGY

AND

ORTHOËPY:

AN ELEMENTARY TREATISE ON PRONUNCIATION

FOR THE USE OF

TEACHERS AND SCHOOLS.

BY

ALBERT SALISBURY, A. M.,

CONDUCTOR OF TEACHERS' INSTITUTES, AND TEACHER OF READING IN
THE WHITEWATER NORMAL SCHOOL, WISCONSIN.

With Physiological Engravings.

WM. J. PARK & CO.
1879.

DAVID ATWOOD, STEROTYPER AND PRINTER,
Madison, Wis. _____

PREFACE.

An accurate and elegant pronunciation forms no small factor of a liberal culture. Careless and uncouth speech is the almost certain index of a general lack of cultivation and refinement.

Orthoëpy, therefore, has rightfully claimed the attention of the student, even in its past estate. But a new interest and an added value have been given to it by the recent rapid development of phonetic science. What once seemed a field of arbitrary custom, is now coming into view as an orderly realm of natural science. Orthoëpy can no longer be studied apart from phonology except by the merest empiricist.

Says Prof. WHITNEY: "The study of phonetics has long been coming forward into more and more prominence as an essential part of the study of language; a thorough understanding of the mode of production of alphabetic sounds, and of their relations to one another as determined by their physical character, has become an indispensable qualification of a linguistic scholar. And he who cannot take to pieces his native utterance, and give a tolerably exact account of every item in it, lacks the true foundation on which everything else should repose."

This little book is submitted to the public in the belief that there is a demand for such a work. It does not aim at any elaborately scientific presentation of the subject treated, but only to give a simple and concise statement of its bare elements, — avoiding, on the one hand, the ancient crudities of statement and nomenclature still current in some quarters and, on the other, the fantastic notions so often projected by eccentric theorizers.

' It has been the resolute purpose of the writer to make a *small* book. It may, perhaps, be thought that he has succeeded too well, since so much has been excluded which would be of interest to the inquiring student. The work lays but slight claim to originality or novelty of matter, and none at all to completeness of treatment.

Though adapted to use as a text-book for classes, it is intended to serve, also, as a teacher's manual, a guide to oral instruction in general exercises or in connection with reading classes; and, lastly, as a *vade mecum* for private reference and study.

The author desires to make acknowledgment to his esteemed colleague, Prof. W. S. JOHNSON, for valuable suggestions in various directions.

STATE NORMAL SCHOOL,
Whitewater, Wis., September, 1879.

PHONOLOGY AND ORTHOËPY.

INTRODUCTORY.

ORTHOËPY is the art of correct pronunciation; it treats of the various sounds of human speech and their proper combination in words.

It also takes account of the notation by which sounds are discriminated to the eye; hence, it is closely related to Orthography, with which it is often confused.

Any thorough or scientific understanding of the facts and principles of Orthoëpy, demands a previous knowledge of so much of Physiology as pertains to the machinery of vocalization and articulation, and so much of Physics as pertains to the theory of sound.

That branch of science which treats of the structure and mode of operation of the bodily organs requisite to speech, is called VOCAL PHYSIOLOGY.

That division of the science of Acoustics which treats of articulate sounds, their physical formation and character, is called PHONETICS, or PHONOLOGY.

The art of representing speech-sounds to the eye, accurately and systematically, giving to each sound some distinct and appropriate symbol, may be called PHONOTYPY.

That part of general Grammar which deals with the current imperfect and but partially phonetic representation of sounds and ideas, is ORTHOGRAPHY.

Orthography is historical in its origin — a matter of growth; Phonotypy is of scientific origin — a matter of theory.

CHAPTER I.

VOCAL PHYSIOLOGY.

1. Organs of Voice:
The organs of voice are in part identical with the organs of respiration. They are the lungs, diaphragm, intercostal muscles, bronchi and trachea, larynx, and pharynx.

2. The lungs constitute the central organ of the vocal machinery. They are two spongy masses enclosed in cases of a tough air-tight membrane called the pleura. These masses are composed of cellular tissue enclosing an immense number of little air-cells, air-tubes, blood-vessels, and nerves.

A clear idea of their general structure may be obtained by examining the lights, or lungs, of any slaughtered animal. The alternate expansion and contraction of the lungs result in the process of breathing, which is the basis of vocalization.

3. The diaphragm is a circular sheet of muscle and tendon which forms the partition between the two great cavities of the body, the thorax, or chest, and the abdomen. In shape, it resembles an inverted basin or low dome, though capable of flattening into the form of an inverted plate or saucer. This muscle is attached to the spine, the lower part of the breast-bone, and the lower ribs all around. The lungs and heart are imme-

diately above it; the stomach and liver, below it. The
fibers of which it is composed radiate from the center,
like the spokes of a wheel.

FIG. 1.—DIAPHRAGM, FRONT VIEW.

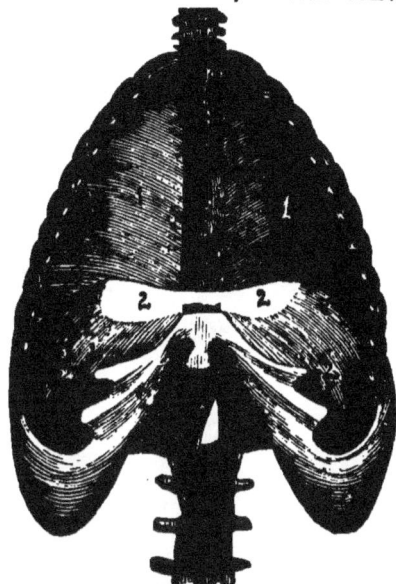

(*From Hooker's New Physiology.*)
1, 1, Cavity of chest.   2, 2, Diaphragm.

4. The intercostal muscles are short, strap-like
muscles, connecting the ribs on either side. The manner of their attachment is shown in Fig. 2. Other
muscles connect the upper pair of ribs to the spinal
column in the neck. By the contraction of all these
muscles, the ribs are elevated at their front extremities,
each pair a little more than the pair above it.

FIG. 2.—ARRANGEMENT OF INTERCOSTAL MUSCLES.

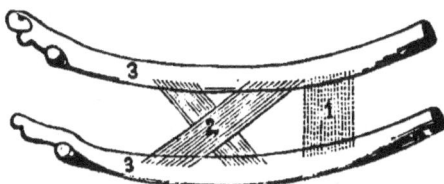

5. The bronchi are minute tubes arising in the air-cells of the lungs and running together to form larger tubes until the last two unite in the trachea.

The trachea, or windpipe, is a flexible tube, composed of rings, of cartilage, or gristle, covered and connected by inside and outside membranes. These cartilaginous rings are incomplete, opening at the back in the manner of a horseshoe, thus allowing the enclosing membranes to sink into a groove, in which the œsophagus, or gullet, partly lies.

The trachea and bronchi form the connecting passage between the lungs and the larynx and mouth.

Fig. 3. — Lung, Bronchi, and Trachea.

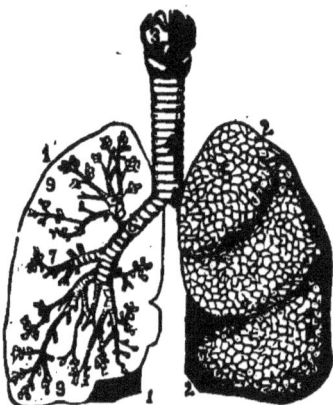

1, Outline of right lung; 2, Left lung; 3, Larynx; 4, Trachea; 5, Lobes of the lung; 6, 7, Bronchi; 9, 9, Air cells.

6. The larynx is an upper story to the windpipe. It is a funnel-shaped tube or box, formed of plates, instead of rings, of cartilage, with enclosing membranes and operative muscles. It is somewhat irregular in shape, the adult male larynx having in front an angular projection known as the Adam's apple.

The cartilages of the larynx are nine in number; of

which the principal ones are the thyroid, the cricoid, the two arytenoid cartilages, and the epiglottis.

The largest of these is the thyroid cartilage, a bent plate forming the front and sides of the shell or case of the larynx, but open behind. This forms the Adam's apple. The cricoid cartilage is so named from its resemblance, in form, to a seal ring. It rests upon the trachea, forming the bed-piece of the larynx. The wide portion or "seal" is at the back, partly filling the posterior opening of the thyroid cartilage. Perched upon the top of this seal and attached to the back of it by strong ligaments are the two arytenoid (ladle-shaped) cartilages. These are movable laterally by several muscles attached to them, and they furnish the rear point of attachment for the vocal chords. The cuneiform (wedge-shaped) cartilages are two minute elastic bodies projecting from the arytenoid cartilages into the folds of the true vocal chords, for about half their length. The cartilages of Wrisberg and of Santorini need not be described here.

FIG. 4.—LARYNX, FRONT VIEW AND SECTION.

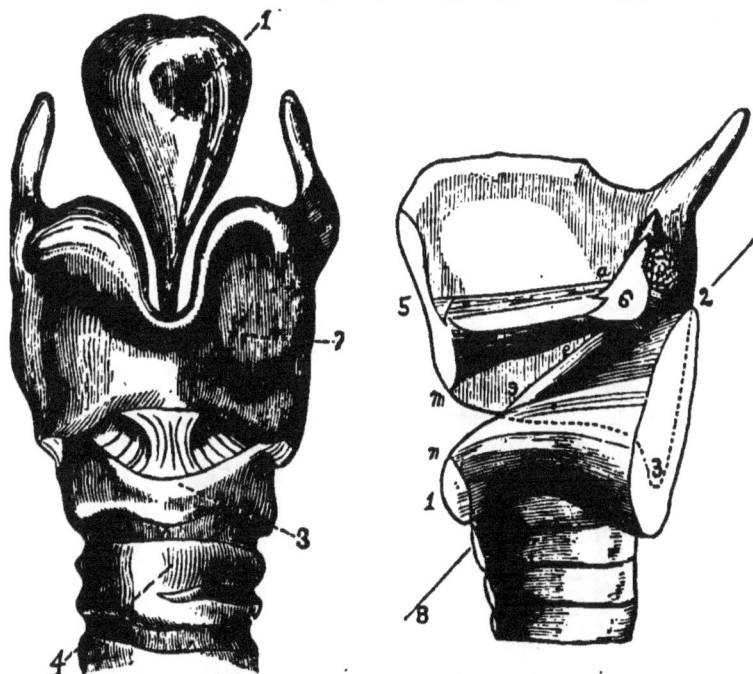

FRONT VIEW: 1, Epiglottis; 2, Thyroid cartilage; 3, Cricoid cartilage; 4, Trachea.

SECTION: 1, 2, Cricoid cartilage; 5, 7, 3, Thyroid cartilage; 6, Arytenoid cartilage; 5, 6, The vocal chords; 9, o, Cricoarytenoid muscle; 8, Trachea.

7. The vocal chords are the special vocal apparatus.
They are situated within the larynx and consist of two
ligaments or bands of fibrous tissue, attached, in front,
to the lower part of the thyroid cartilage and, rearward,
to the two arytenoids. These ligaments, along with
certain muscles, are inclosed in two folds in the lining
membrane of the larynx. When inactive, as in ordinary
breathing or whispering, they present to the eye the ap-
pearance of two rounded ridges or cushions on the sides
of the passage or cavity of the larynx.

> Above these are two other somewhat similar folds, known as
> the false vocal chords. Their function is not certainly known.
> Between the true and false vocal chords, are two lateral depres-
> sions or cavities called ventricles.

8. The glottis is the aperture, or opening, between
the vocal chords. When the chords are at rest the glottis has somewhat the shape of a key-hole; when they are
active, as in vocalization, the opening diminishes to a
mere line.

FIG. 5. — VOCAL CHORDS AND GLOTTIS.

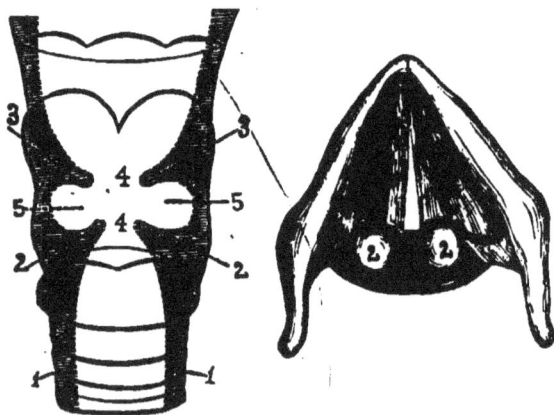

SECTION OF LARYNX: 1, Trachea; 2, The true vocal chords; 3, The
    false vocal chords; 4, 4, The glottis; 5, 5, Ventricles.
GLOTTIS: 2, 2, Arytenoid cartilages.

9. The epiglottis is a lid or valve formed to shut down over the glottis in the act of swallowing. Though enumerated with the cartilages above, it is in part of tendinous tissue and may be called fibro-cartilaginous.

The masticated food slides over the upper surface of the epiglottis into the œsophagus. If, by reason of tardy or imperfect closure, the food passes into the larynx instead, a convulsion follows; and we say that we have "swallowed the wrong way."

10. The pharynx is a sort of chamber between the mouth and the larynx. It may be compared to an inverted sack with several openings in one side. It opens downward into the larynx and œsophagus; forward, into the mouth, the nasal passages, and, by the Eustachian tubes into the drum of the ear.

The pharynx, with the mouth and other cavities of the head, performs the office of a resonator or tone-magnifier, giving greater power and richness to the tones of the larynx.

THE PROCESS OF BREATHING.

11. Inspiration, or inhalation, is the process of taking breath. In order to inhale, the cavity of the chest is enlarged, thus tending to create a vacuum around the pleura, or lung-case. The resistance being thus removed, the outside air *falls* through the trachea and bronchi into the lung-cells, thereby causing the lungs to expand and follow up the walls of the chest.

This enlargement of the chest is produced by a double agency: (1) The diaphragm is depressed, or flattened, partly by contraction of its radiating fibers, and partly by an outward movement of the walls of the abdomen, to which its outer rim is attached, thus enlarging the chest downward. (2) The ribs are moved upward and outward by the contraction of the intercostal and pectoral muscles, and the chest is thus enlarged upward.

The first-named agency, the action of the diaphragm, is the one which should be most relied on and cultivated for all vocal purposes, and those of general health as well. It is impossible to over-estimate the value of a full and proper use of the diaphragm. Females are especially prone, through improper dressing or other bad habits, to err in the disuse of this organ. Feeble health and feeble voices are but the natural result.

12. Expiration, or exhalation, is the opposite of inspiration. When the various organs have completed the movements of inspiration, they reäct by their own elastic force. This reäction is aided by that of other organs, as the intestines, that have been crowded upon, and by the weight of the ribs; and the air is driven or squeezed out of the lungs.

THE PROCESS OF VOCALIZATION.

13. In ordinary respiration the vocal chords lie relaxed and flattened against the walls of the larynx; and only a slight rustling sound, if any, is produced by the friction of the air breathed out.

Vocalization, or the production of voice, is accomplished in the following manner: By the contraction of the proper muscles in the larynx, the two arytenoid cartilages, sitting on the back margin of the cricoid, are moved towards each other, thus bringing the vocal chords nearer together and narrowing the glottis to a mere chink. At the same time, the thyroid cartilage is drawn downward and slightly forward, thereby tightening the chords. The outward current of breath, driven against and between the now tense folds of membrane, sets them into a more or less rapid vibration, somewhat similar to that of the reeds in an accordion. This vibration is communicated to the confined column of air, as by the reed of a clarionet; and the air-waves, thus set in motion, are strengthened by the pharynx and ultimately affect the ear of the hearer. The result of all this is a vocal tone more or less pure, or in other words, voice.

The action of the larynx is compared to that of a reed instrument. In fact it combines the three principles on which all musical instruments are constructed; the string, the reed, and the vibrating column of air as in the flute.

THE ORGANS OF SPEECH.

14. The Organs of Speech are those organs which are employed in modifying the breath, vocalized or unvocalized, for the purpose of expressing thought.

They are the tongue, lips, palate, teeth, and nasal passages.

By various combinations with each other, they obstruct the outward movement of the breath from the pharynx, and so give rise to a great variety of modifications of the natural or fundamental tone of the voice.

In whispering, unvocalized breath is modified or affected by these organs to suit the purposes of speech.

15. The tongue is not the simple paddle-shaped organ which it is commonly supposed to be from observation of its upper surface, but rather a thick *cushion-shaped* mass of muscular fibers in apparently complete confusion, but really so disposed as to be capable of producing motion in any and every direction or several directions at once.

In phonology it is considered, for convenience of description, as having three parts, the tip, the front or blade, and the base.

16. The palate is the roof of the mouth. The fixed front portion is called the *hard palate*. Continuous with it, backward, is a yielding muscular and membranous awning, separating the mouth from the nasal passages and the upper part of the pharynx. This is the *soft palate*. Dependent from this is a conical appendage called the *Uvula*.

The soft palate is capable of depression and other movements.

17. The nasal passages admit of closure at their inner extremities by the action of the soft palate. The presence or absence of this closure is very essential to the production of certain sounds.

The lips and **teeth** need no description. The former are of great importance in articulation; the latter, of but little.

FIG. 6. — SECTION OF HEAD, SHOWING TONGUE, ETC.

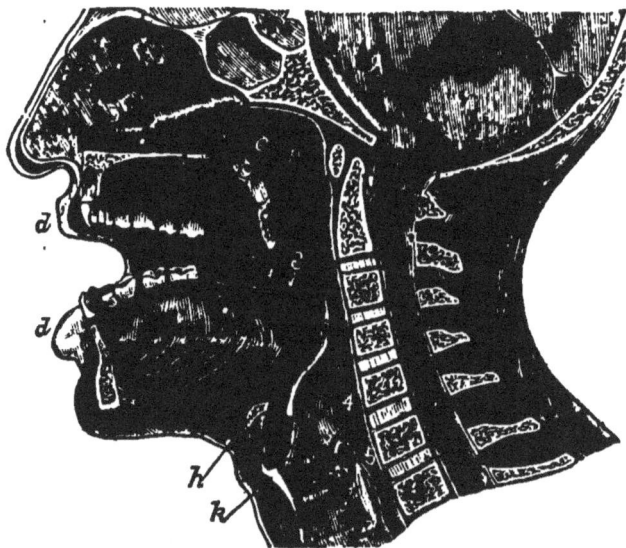

b, Tongue; *c*, Section of palate; *d, d*, Lips; P, Pharynx; S, Epiglottis; *u*, Uvula; V, Glottis; 5, Passage into œsophagus; *h*, Hyoid bone; *k*, Thyroid cartilage.

THE PROCESS OF ARTICULATION.

18. The distinctive and crowning process of speech is that of articulation, a process as complex and intricate as it is essential.

The tongue, by its power of manifold motion, moves forward and back, narrows and widens, arches and flattens in its several parts; the lips open and contract; the palate rises and lowers; the nasal passages are closed and unclosed; the teeth approach and separate,— all these movements take place in every varying combination, shaping the column of vibrating breath; and from each separate combination results a sound of distinct and recognizable quality, capable of appropriation as a thought-symbol.

The subject of articulation is further discussed in a succeeding chapter.

THE ORGANS OF HEARING.

19. The ear is not an organ for the production of voice, but its receiving instrument.

Sound-waves in the air, or other medium, are focused by the external ear upon the tympanum, a cavity covered by a thin membrane similar in its arrangement and function to the head of a drum. A number of small bones in contact with the inner side of this membrane transmit the vibrations to the internal ear, whence the auditory nerves communicate with the brain.

20. Summary. The diaphragm and other muscles, by their alternate movements, operate the lungs. The breath, forced from the lungs, passes through the bronchi and trachea into the larynx. The vocal chords, when tensely drawn across the cavity of the larynx, set the column of breath into vibration. This vibration, increased by the resonating action of the pharynx and other cavities, is communicated to the external air, and at length falls as a tone upon the listening ear.

CHAPTER II.

PHONOLOGY.

21. Phonology, or **phonetics,** is the science of articulate sounds, and treats of their physical character and formation.

It is a branch of the science of acoustics.

22. Sound is the effect produced upon the auditory nerve by vibrations of the air or other conducting media.

Water and solid substances, as wood, or metal, are good conductors or media of sound-waves; but usually, if not always, a greater or less portion of air enters into the chain of communication.

23. Sounds are classified as tones and noises.

A tone is a sound produced by regular, or periodical, vibrations of the sounding body. It admits of uniform continuation, and is usually agreeable to the ear.

A noise, is a sound produced by irregular, or non-periodical, vibrations — the motions of the sounding body changing irregularly.

A combination or co-incidence of discordant tones, as when the keys of a piano are all struck at once, is also a noise.

A water-fall, for instance, or a machine in motion, has its uniform tone, or key-note, usually, however, rendered almost unnoticeable by the multitude of discordant noises — splashings, thumpings, etc. — which accompany and overpower it.

24. Voice is tone produced by the mutual action of the larynx and the breath from the lungs.

It is, perhaps, possible, though exceedingly uncommon and unnatural, to produce voice with the in-going breath.
The pure, unmixed, unobstructed product of the larynx is the sound heard in the English word *ah* when clearly uttered. It

is the same in all persons without distinction of age, sex, or race. It is capable, however, of extensive variation in pitch, this being the sole modification of voice which the unaided larynx can effect.

The volume, or quantity, of voice depends upon the amount and the rate of expulsion of the out-going breath. It is controlled chiefly by the diaphragm and the abdominal muscles.

25. Speech is either voice or breath modified, by articulation, for the purpose of expressing thought.

Singing without words, the wailing of an infant, etc., are examples of voice without speech. Ordinary whispering is speech without voice.

Common speech employs a mixture of vocalized and unvocalized breath duly articulated, a combination of tones and noises.

26. An oral element, or **elementary sound,** is, strictly speaking, a sound of human speech which cannot be analyzed, or separated into parts. It is produced with a single and fixed position of the organs of speech.

In common speech, however, the term has been loosely applied also to certain couplets or combinations of sounds, as the diphthongs. This leads to the expression, compound element, a contradiction in terms, but too firmly established by usage, perhaps, to be abolished.

27. The number of oral elements, including compounds according to the popular usage above mentioned, is given by Webster's Dictionary as *forty-five*. The number recognized by Worcester is practically the same.

Phonetists have not been able to come to any agreement, as yet, in regard to the exact number of distinct and true elements in our language.

The number of possible speech sounds is almost infinite. Alexander Ellis, the great English phonologist, has invented a notation for about 400 of them, which he calls the Palæotype.

28. Classification. The oral elements admit of classification in several different ways or modes, varying according to the basis of classification employed.

The most familiar classification is that into vocals, sub-vocals, and aspirates.

A vocal, or **vowel-sound,** is a tone of the voice but

2

little or not at all modified, or interrupted, by the organs of speech.

A sub-vocal is a tone of the voice greatly modified,
or interrupted, by the organs of speech.

An aspirate is a mere breathing more or less modified by the organs of speech.

> Vocals, sub-vocals, and aspirates, are also called, with great
> propriety, tonics, sub-tonics and atonics.
> Vocals, or tonics, are vocal tones nearly pure, i.e., but little mixed
> with mere noise. Sub-vocals, or sub-tonics, are impure tones,
> or tones so greatly mixed with noise, the rustling of breath
> against the organs, etc., that the noise predominates over the
> tone more or less. The tone is covered by the noise and be-
> comes *under*tone.
> Aspirates, or atonics, contain no vocal tone, being produced
> with the vocal chords in a state of inaction.

It will thus be seen that this classification is based
upon the amount of vocal tone — much, little, or
none — which the sound contains.

29. A vowel is a letter used ordinarily to represent
a vocal, or tonic, sound.

A consonant is a letter used ordinarily to represent
a sub-vocal or an aspirate sound.

> Loose popular usage, it is true, employs the term vowel to de-
> note a vocal, or tonic, sound; but it is needful for scientific pur-
> poses to restrict the meaning of the word.
> Nothing can be phonetically more absurd than the ancient and
> still common definition of a consonant, as "a sound which can-
> not be uttered without the aid of a vowel." There is no sound
> in our language which cannot be uttered independently and
> alone. Nor is the later one, "a sound which cannot be uttered
> without bringing the organs into contact," much better. It
> is true of only a part, at most, of the consonant sounds. And it
> is always better to confine the word consonant, as a noun, to the
> indication of a class of letters.
> The English vowels are *a, e, i, o, u,* and sometimes *y*. A more
> scientific statement would add to *y* also, *l, n,* and *r*. *W* is never
> a vowel.

30. Cognates *(cog,* with; *natus,* born) are those pairs
of consonant sounds, one sub-vocal and one aspirate,
which are produced with the organs of speech in the

same, or very nearly the same, position for both, as *b* and *p*, *v* and *f*.

A table showing all the cognates will be found in the chapter on Orthoëpy.

31. A diphthong *(di,* double; *phthongos,* voice) is a combination of two vocals, or vowel-sounds, in one utterance or syllable. It may be represented to the eye by two letters or one.

The essential characteristic and test of a diphthong is that it requires a change in the position of the organs of speech during the continuance of a tone.

There are six diphthongs in the English language, as heard in the words *out, oil, ice, use, oh, ate.*

The old distinction of ."proper" and "improper" diphthongs, is essentially absurd and mischievous; and there is no such thing in the English language as a "triphthong."

32. A digraph is a combination of two letters to represent one sound.

These letters may be vowels or consonants; hence we may have vowel digraphs, as *ai* in *said,* or consonant digraphs, as *ph* in *phiz.*

A trigraph is a combination of three letters to represent one sound or a diphthong, as *sch* in *schist, eau* in *beau.*

The terms digraph and trigraph, like vowel and consonant, might be considered as belonging to Orthography, but they are also necessary here.

33. Another classification of the oral elements is that based upon the *kind* of modification which the sounds receive, that is, upon the special organs of speech used in forming them.

The several classes take their names from the organ most prominently in use. .

A labial is a speech-sound modified chiefly by the lips, as the sounds of *o, b,* and *p.*

A palatal is a sound modified chiefly by the palate, as the sounds of *e, g,* and *k.*

A lingual is a sound modified chiefly by the tongue, as the sounds of *l*, *d*, and *t*.

The lips, being two and external, are more independent than other organs in their action. The tongue and palate assist each other, the sound being named, or classed, according to the greater prominence of either organ in the work of modification. The teeth also assist in the formation of certain sounds, which may therefore be called labio-dentals, linguo-dentals, etc.

Sounds which owe their peculiar quality in part to an openness of the nasal passages, are called nasals, as the sounds of *m* and *n*.

34. **Long** and **short** are terms which apply only to vocals. Vowel sounds differ from sub-vocals in that they are less interrupted by the organs of speech. They differ from each other in *quantity* or duration, and in *quality*. With reference to quantity, they are classified as *long* and *short*.

Long vowel sounds are those which may be, and usually are, prolonged in their utterance, as *a* in *pay*, *oo* in *woo*, etc.

Short vowel sounds are those which, in ordinary speech, do not admit of prolongation, as *i* in *fit*, *o* in *not*, etc.

They are in the English language peculiarly abrupt or "explosive " in their utterance. The prolonging of a short sound results in "drawling."

35. Each vowel has a "regular" long and short sound which it in most cases represents, and one or more "occasional " or irregular sounds. The regular long and short sounds of a given vowel, in English, are not necessarily, nor even usually, the natural correlatives of each other.

Long Sounds.		*Short Sounds.*	
a, as in	fame.	*a*, as in	fat.
a, "	father.	*a*, "	ask.
a, "	fall.	*e*, "	met.
a, "	fare.	*i*, "	pit.
e, "	mete.	*o*, "	coffee.

Long Sounds.			*Short Sounds.*		
e, as in	-	verse.	*oo,* as in	-	foot.
i, "	- -	mine.	*u,* "	- -	up.
o, "	-	tone.			
oo, "	- -	boot.			
u, "	-	tune.			
u, "	- -	urge.			
ou, "	-	sour.			
oi, "	- -	oil.			

CORRELATIVE LONG AND SHORT SOUNDS.

Long.						*Short.*
a in care	-	-	-	-		*e* in met.
a in father	-	-	-	-		*a* in ask.
a in fall	-	-	-	-		*o* in coffee, on.
e in mete	-	-	-	-		*i* in mit.
oo in boot	-	-	-	-		*oo* in foot.
u in urge	-	-	-	-		*u* in up.

The sounds of *a* in *fat, e* in *verse,* and the six diphthongs, have no English correlatives.

36. Quality. Vowel sounds differ in quality according to the different positions of the organs during their utterance, every new adjustment of the organs producing a distinct effect upon the ear.

The various terms, as *flat, grave, broad, obtuse,* etc., which have been used to indicate quality of sounds, are rather misleading than useful.

The study and discrimination of the nicer and more difficult shades of sound, and of the configurations by which they are produced, is a matter of much importance to scientific students of language.

37. Semi-vowels are those sounds which, in their degree of modification, stand on the border line between vocals and sub-vocals, and are thus capable of use in either class. They are the sounds of *w, y, l, n,* and *r,* and perhaps even that of *m.* See section 60.

The term "semi-vowel" is not extremely accurate; but for the want of a better is used here, and is likely to continue in use, with a more scientific application than formerly.

38. For purposes of description all the classifications outlined in the preceding section are useful. Other classifications according to physiological character have been made, but that of Prof. W. D. Whitney — adapted, in this work, to the notation of Webster's Dictionary — will be chiefly adhered to, as on the whole the simplest and most satisfactory for practical purposes.

The following diagram presents this classification to the eye at a single view:

39. Diagram of the oral elements.

(Classified according to Mode of Formation.)

The above diagram is, in the main, self-explanatory to one acquainted with Webster's notation.

Starting with the pure, open tone *ah*, the sounds in each series are arranged in the order of openness, downward direction indicating increase of closure. Diphthongs are represented upon curved lines connecting their component elements.

Consonant sounds not belonging fully to any one of the three series, are placed beside that to which they are the most closely related.

If the student will produce, successively, all the sounds of each series in the downward order, he cannot fail to observe the gradual and uninterrupted closure of the organs concerned.

DESCRIPTION OF THE ENGLISH SOUNDS.

A brief description of each of the sounds recognized by the English dictionaries is here given as possibly the most practical and serviceable part of this treatise. It is thought best to take them up in the order in which they naturally occur in the several series as exhibited in the preceding diagram (Sec. 39).

THE VOWEL SOUNDS.

40. A as in ah, far. *Italian a.* This is the fundamental tone of the human voice, the pure product of the vocal organs. Its proper production requires an extreme openness of the organs of speech, allowing the column of fully vocalized breath to pass without obstruction at any point. All other vocals, and the sub-vocals, may be considered as simply modifications of this tone.

This noblest of sounds has become, unhappily, too rare in our language, constituting at present less than one-half of one per cent. of our whole utterance, as against five per cent. in the German and *thirty* per cent. in the ancient Sanscrit.

41. A as in ask. *Short Italian a of Webster — Intermediate a of Worcester.* This sound differs from the preceding one in quantity, being "short" or explosive. When perfectly produced, it requires the same extreme openness of the organs as the full *Italian ä;* but it is liable, even in the mouths of good speakers, to a slight modification by closure. In instruction, however, the full openness should be insisted upon.

Uneducated speakers often use, in place of this elegant sound, in words like *dance, grass*, etc., a corruption or drawling of the *short a*, a coarse and most disagreeable error.

LABIALS.

42. O as in on, coffee. A as in what. *Short o, (short broad a).* This sound so closely resembles the *short Italian a*, as to be very often confounded with it. Its proper utterance requires that the column of vocalized breath should be slightly obstructed by contraction of the lips, drawing the corners of the mouth slightly towards each other. The sound closely resembles that of *a* in *fall*, but is short, or explosive, admitting of no prolongation.

> Much care should be taken with this sound; for, while it is one of the finest in the language, it is probably the most abused — the pronunciation of such words as *not*, *what*, *on*, *hog*, *fog*, *watch*, etc., with the sound of *Italian a*, more or less shortened, being the invariable custom of the majority of people in some localities, especially in the Northwestern States. This is a provincialism which deserves no toleration.

43. A as in awe. O as in or. *Broad a.* Broad *a* resembles *Italian a* in quantity, being long; but it is modified by a contraction and consequent projection of the lips, which lengthens the cavity of resonance.

The position of the lips is a trifle closer than that for *short o*, from which sound this differs but slightly, except in duration.

> A few words like *cross*, *cost*, *salt*, are often pronounced with a quantity intermediate between the regular *short o* and *broad a.* This distinction need not be insisted upon, however.

44. O as in ho. *Long o.* This sound is a labial diphthong. It begins with a position of the lips somewhat closer than that for *broad a*, which position is still further closed during the continuance of the tone, which vanishes in the sound of *oo* as in *coo*.

> In unaccented syllables ending in *o*, and in a few words like *wholly*, the vanish is omitted, and only the radical or brief initial sound remains.

45. OO as in foot. O as in wolf. U as in put. *Short oo — (u medial).* A still closer lip position than the one for beginning *long o*, with an explosive emission of voice, gives the short vocal known as *short oo*, heard in *foot*, *push*.

46. OO as in boot. O as in do. U as in rude. *Long oo — (slender o).* A prolongation of the vocal tone with a slightly closer position than for the preceding sound *(short oo)*, yields the sound heard in *coo, do*. It is the closest of the labial vocals, the next stage of lip-closure resulting in the sound of *w*.

> This sound is often indolently contracted into the *short oo* in such words as *broom, room, soon*, and even *food*. To pronounce these with the sound of *oo* in *foot* is grossly negligent, though only too common.

47. Ou as in sound. Ow as in cow. If the lips change from extreme openness, as in *ah*, to the extreme closeness of *oo* while the vocal tone continues, the result will be the labial dipthong, *ou*, which, therefore, may be considered as the sum of the whole series just discussed. Its analysis is $ou = \breve{a} + \infty$.

> A caution is here necessary. Many speakers begin this diph-thong with the sound of *short a*, as *caou* for *cow*. This vulgar error is usually thought to be a peculiarity of illiterate " Yankees," but it is by no means limited to New England.

PALATALS.

> This series of sounds might with greater accuracy be termed Linguo-Palatals, since the part played by the tongue is so great; but the simpler term has the sanction of high authority, at least.

48. A as in at. *Short a.* This simple and familiar sound differs but little in position from the *short Italian a*, though quite distinct to the ear. A slightly different adjustment of the soft palate from that for *ah*, and a slight lifting of the blade of the tongue, constitute its peculiarity.

> Like other short sounds it should receive a neat and elegant utter-ance, any prolongation of it destroying its true character.

49. E as in met. *Short e.* A still closer approach of tongue and palate than that required for *short a*, is necessary for the production of *short e*, the tongue being thrust well forward, and its middle portion considerably arched.

> The only caution needful is that against prolongation in speaking. It may be prolonged in singing, however.

50. A as in care. E as in there. *Circumflex a,* *(Worcester's a long before r).* This sound has been thought by some to be identical with the preceding one, *short e.* It differs from it, when correctly uttered, in being somewhat closer and in admitting of moderate prolongation. It occurs, in accented syllables, only before the sound of *r*, and has by some been erroneously regarded as a diphthong, owing to the semi-vowel character of the *r* itself. It is a simple element, and constitutes the radical, or initial, part of the diphthong, *long a*, heard in *pay.*

The majority of American speakers—in the interior, at least— give in place of this sound one of somewhat different character. It may be described as *short a* drawled or prolonged. This practice receives a sort of left-handed sanction from Webster, " provided it be given without undue coarseness or breadth; " but it is usually avoided by cultivated speakers. There is something to be said in its favor scientifically, however, as furnishing the correlative long sound, otherwise missing, of *short a.* With this utterance, it should be placed above *short e* in the scale (Sec. 39).

51. A as in pay. E as in prey. *Long a. Long a* is a linguo-palatal diphthong. It begins with the preceding sound in the series, *a* as in *care*, and closes with the sound of *e* in *me.* This involves a considerable closure of the palate and tongue during the utterance of the sound.

In the utterance of *a* in *care*, the tongue is immediately drawn back and narrowed to form the palatal *r;* but in forming *long a* the tongue is pressed still further forward, and is crowded against the upper teeth to form the vanishing element. In unaccented syllables, the vanish is sometimes omitted.

52. I as in it. Y as in abyss. *Short i.* This sound most resembles that of *e* in *me.* It is slightly more open in its formation, being, in closeness, midway between the radical and the vanishing parts of *long a.* It is a true abrupt or short sound; and even when prolonged, it is still distinct from *long e.*

53. E as in me. I as in pique. *Long e.* This is
the closest of the palatal vocals, the next stage of palatal closure yielding the semi-vowel *y*, as in *yet*. For the
formation of *long e*, the edges of the tongue are pressed
against the teeth, while its middle portion is almost in
contact with the palate throughout its whole length,
thus leaving a very thin passage for the breath.

54. I as in ice. Y as in my. *Long i.* This is a
palatal diphthong. For its production, the tongue and
palate are placed in the extreme open position of *Italian
a* and closed, during vocalization, to the extreme close
position of *long e*. Thus, like *ou*, it is the sum of a series
of sounds. Its analysis is usually given as $\bar{\imath} = \breve{a} + \bar{e}$,
which is practically correct.

A common fault in its utterance consists in not commencing with
a sufficiently open position of the mouth.

MIXED DIPHTHONGS.

Ou and long *o* are labial diphthongs; *long i* and *long a* are palatal
diphthongs. Two others combine the action of all the organs
in such a way as to entitle them to the name, mixed diphthongs.

55. U as in use, tune. Ew as in new. *Long u.*
The diphthong *long u* presents two distinct phases to
the ear, as heard in the words *use* and *tune*. If the
palate and tongue be placed in the close and tense position required for the sound of *y* in *yet*, and then opened
while the lips close to the position of *oo* in *woo*, the resulting voice-sound will be that of *long u* at the beginning of a syllable, as in *union, use*, etc. In any other
place than the beginning of a syllable, however, it is
almost impossible to perfectly form the *y* sound; hence
a more open position is substituted, that of *short i*, as
in *it;* and the *u* becomes a combination of *short i* and
long oo, the *ĭ* accented, but very quickly uttered. But
for this change from *y* to *short i*, the words *tune* and
duke would become, in most mouths, *choon* and *jook*.

Long u is one of the most difficult and trying sounds of our language. Its analysis may be represented thus,

$$\mathfrak{u} = \left\{ \begin{matrix} y \\ i \end{matrix} \right\} + \infty.$$

56. Oi as in oil. Oy as in boy. Its position is that for *broad a, awe*, changing to that of the close palatal, *short i; oi = ô + ɩ.*

LINGUALS.

The linguals differ from the palatals in the relative prominence of the tongue as a modifying instrument. This is more plainly seen in the consonant than in the vowel sounds of the series.

57. U as in up. O as in son. *Short u.* This is an open sound, being like *short a* and *short o* but one remove from *Italian a*, though in a different direction. The slight closure necessary to transform *short Italian a* into *short u*, is effected by a slight elevation of the base of the tongue.

The sound is one of easy utterance, requiring little muscular effort, and therefore liable to intrude itself into many places where it does not belong, to the exclusion of more elegant sounds — especially in unaccented syllables. The excessive use of it is a mark of laziness and barbarous negligence in speech.

58. U in urge. O in word. *Circumflex u.* A slightly greater elevation of the back part of the tongue toward the soft palate than that for *short u*, with prolongation of the tone, gives the sound of *u* heard before *r* final or *r* followed by another consonant. It is a comparatively open sound, and easy of utterance, differing from *short u*, to the ear, chiefly in its greater duration.

59. E as in verse. I as in girl. *Tilde e.* This is a close lingual sound, the tongue being well raised in all its forward part, while the teeth are brought nearer together than for the preceding sound *(u in urge)*. It has been described as an intermediate between *short e* as *merry*, and the *u* in *urge:* though it is commonly confused by great numbers of the people with the latter sound. The distinction between the two is insisted upon by such authorities as Webster and Smart.

The direction to be given to students is: Keep a close position of all the organs and form the sound well *forward* in the mouth. The *u* sound can be made with an open mouth, this cannot.

This seems to be identical with the German *umlaut o*, as in *Goethe*. It is also quite similar to our lingual *r*, which accounts for a part of our difficulty with it, and for the fact that the word *Goethe* is so often sunk into " *Gerty* " in pronunciation.

In trying to escape confounding this sound with *u* in *urge*, we are in danger, also, of going to the opposite extreme of making it too closely resemble *short e*.

SEMI-VOWELS.

60. As the difference between vocals and sub-vocals is only a difference in degree of modification or obstruction by the organs of speech, it is but natural that there should be a stage of uncertainty, a sort of border-land, between them. Hence some writers, with much reason, recognize those sounds which lie along this border-line as a separate class, under the name of semi-vowels.

61. W. The labial semi-vowel is represented in English by the letter *w*. It is formed by a lip-closure so extreme as to lessen the purity of the tone considerably below that of *long oo*, though not so far as to prevent prolongation of the sound.

62. Y. The palatal semi-vowel is the sound of *y* in *yet*, which bears the same close relation to *long e* that *w* does to *long oo*. The position of the organs is similar to that for *long e*, but one degree closer, reducing the tone to a mere buzz or hum.

The tongue is slightly drawn back from the *ē* position, and the pressure against the teeth is increased.

63. R. Closely related to the sound of *e* in *her (tilde e)*, are the two sounds of *r*. The lingual *r*, heard at the beginning — or anywhere before the vowel sound — of a syllable, is formed by placing the tongue well forward and turned upward so that the breath is passed over its extreme tip, producing a very slight trill or vibration. The position differs from that of *tilde e* in the turning up of the tip of the tongue.

The palatal or uvular *r*, heard at the end of a syllable, or whenever not immediately followed by a vowel, as in *far, farm*, can be produced without the aid of the tip of the tongue, being formed farther back in the mouth.

> This is clearly a different sound from the lingual *r*, but the two are not discriminated by some ears. The common and disagreeable error of failing to sound the palatal *r* — giving *fahmah* for farmer, etc., is usually taken as an evidence of affectation. It is often, however, a matter of innocent, ignorant habit rather than affectation.

64. L. The sound of *l* is of about the same closeness as the lingual *r*, the tip of the tongue, however, being placed against the upper teeth or the roof of the mouth, and the breath allowed to escape over the edges of the tongue.

> It is the semi-vowel character of *l* which allows it to become the vocal basis of a syllable, as in *able*, *shovel*, etc., in which the *e* is entirely mute, and yet the words are dissyllables.
> The substitution of *l* for *r* by Chinamen is doubtless a consequence of the similarity of the two in degree of interruption.

65. N. The nasal sound of *n*, in *nail*, has the same peculiarity as the foregoing, often constituting a syllable of itself, as in *heaven*, *cotton*, where the preceding vowel is silent.

In the production of this sound, the tongue is placed against the hard palate in such a way as to wholly obstruct the oral passage, the breath escaping through the nasal passages instead.

SEMI-VOWELS. { Labial, *W*.
{ Palatal, *Y*.
{ Lingual, *R*, *L*, *N*.

OTHER SUB-VOCALS.

66. As already defined, sub-vocals are tones produced in the larynx, but greatly modified in the mouth. They are thus impure tones, or, as the name implies, undertones. Vocals are also subject to obstruction. as we have seen, but not to the same degree.

The obstruction of the breath gives rise to friction and a mingling of mere noise with the tone. When this admixture of noise reaches such a degree as to predominate over and partially obscure the tone, the sound is called *sub*-vocal.

LABIALS.

67. V. If the edges of the upper teeth be placed upon the lower lip, and the vocalized breath forced between the teeth, the sound of the letter *v* will be produced. This sound would be more correctly named labio-dental.

68. M. Let the lips be closed entirely and the vocalized breath be allowed to pass only through the nose. The resulting sound is that of *m*. It differs from that of *n* only in its initial quality and not in its continuation.

This sound is sometimes ranked among the semi-vowels, since it is possible for it to serve as the vowel element of a syllable, as in the common contraction *yes'm*, and the ejaculation *m'h'm*. These are hardly legitimate words, however.

69. B. If, now, the nasal passages be covered by the soft palate, while the action of the larynx continues, we have the sound of the letter *b*, a sound requiring complete contact of the organs and, so, not capable of prolongation.

The common error in its separate production consists in allowing the lips to part, thus producing not the sound of *b* alone, but in connection with a neutral vowel — a combination best represented by the syllable *buh*.

PALATALS.

70. Zh. The sound usually represented by *z* before *long u*, as in *azure*, or by *z* or *s* before *i*, as in *osier*, is produced with the blade of the tongue in close proximity to the hard palate and the teeth shut, or nearly so.

It is a simple element, produced without change of position, though tongue, teeth and palate conjoin in its formation. It is thus not a pure but a mixed palatal.

This sound has been treated as a compound of *z* and *y*, but the fact seems to be that the utterance of *z* and *y* in succession is impossible without a hiatus, and this element, somewhat similar to them both, is *substituted* for them. Though known as the sound of *zh*, it is never represented by that combination of letters, which, indeed, does not occur in the English language. The sound might, with more propriety, be called the second sound of *z*.

71. J. The sound of *j* is also a mixed palatal. It has generally been considered a compound of the sound of *d* and the one just discussed, *zh*. It is undoubtedly a compound, a sub-vocal diphthong, so to speak; but the analysis mentioned, *d+zh*, is of doubtful accuracy. *D+y* would seem to be nearer the truth; but the second element is, in all probability, a sound which does not occur separately in our language. The sound of *j* differs from that of *zh* in the still greater elevation of tho tongue, forming a temporary contact with the hard palate, which is then suddenly broken, the closed teeth parting at the same instant and allowing the breath to escape forcibly.

When *j* is uttered without a vowel immediately ensuing, it is inevitably followed, or closed, by the sound of its aspirate cognate, *ch*.

72. Ng, N. The second or palatal sound of *n*, usually called *ng*, is produced by bringing the soft palate and the back part of the tongue into complete contact, compelling the breath to escape through the nose, as in *m* and *n*.

From being produced so far back, it is often called a guttural, or throat sound. It is very often displaced by the first or common sound of *n* in the mouths of indolent or negligent speakers. Though often represented by the digraph *ng*, it is frequently represented by *n* alone, as in *fin-ger*, *lin-ger*, etc., in which words the *g* has its own sound and forms no part of the representation of this sound.

73. G. With the base of the tongue and the soft palate in perfect contact, close also the nasal passages; the attempt to vocalize will result in the sound of *g*, as in *gate*, which occupies the same place in the palatal series as *b* in the labial — the last, or closest, sub-vocal.

The letter *g* is unfortunately often used to represent the more open sound of *j*.

LINGUALS.

74. Th. The sub-vocal *th*, as in *this*, is a linguo-dental. Though the occasion of so much trouble to foreigners learning our language, it is of the easiest production.

Place the tip of the tongue under and against the edges of the upper teeth and expel the vocalized breath between the teeth.

The above simple direction, aided by reasonable attention and perseverance will enable any person, whose mouth has not become actually ossified, to acquire this sound perfectly.

75. Z. To produce the sound of *z*, as in *buzz*, the tongue takes the same general position as for the trilled *r;* but the tip is a little less elevated and is brought very near to the teeth, which are nearly or quite closed. ·

The close resemblance of this position to that for *th*, accounts for the Frenchman's treatment of that sound, *th*, in speaking English.

76. D. Place the tip of the tongue against the hard palate, so as to completely obstruct the oral passage, the position for *n;* close the nasal passages also, permitting no breath to escape. The attempt to vocalize will then

3

result in the sound of the letter *d*, the last sub-vocal in the lingual series. Like *b* and *g*, it is non-continuant.

> The same error is made in attempting to produce this sound sepa-rately that was mentioned in connection with *b*. The organs are allowed to part, permitting the breath to escape and form the syllable *duh*. Let no breath escape ·until the tone has ceased.

ASPIRATES.

77. Wh. If unvocalized breath be expelled with the lips closely contracted, as for the semi-vowel *w*, the sound produced is that represented by *wh*, as in *what*, a labial aspirate.

> It has been a disputed point whether this sound is simply a whis-pered *w* or a compound, *h* + *w*, the *w* of the compound being the full sub-vocal. The editors of Webster have seemed to waver on this point, but such phonologists as Ellis and Bell pronounce it to be a distinct and simple aspirate.
> A failure to discriminate between this sound and its cognate, *w*, constitutes one of the peculiarities of the English cockney dia-lect, in which *when*, *what*, *which*, become *wen*, *wat*, *wich*, etc.

78. F (Ph). The labio-dental aspirate *f*, is the cog-nate of *v*. The lower lip is placed against the upper teeth and unvocalized breath expelled.

79. P. If the unvocalized breath be accumulated behind the closed lips and they be suddenly parted, the puff of escaping air yields the sound of the letter *p*, a labial aspirate.

80. H. The forcible aspiration known as the sound of *h*, is usually classified as a palatal. It is, however, somewhat anomalous in character, being capable of pro-duction in any of ·the vowel positions indifferently, as can be seen by uttering the words *aha*, *oho*, and similar combinations, in which the whole is pronounced with-out change of position.

> It is simply a sudden expulsion of the breath with any open posi-tion of the organs, and the Greeks were consistent in rejecting it as an independent sound, and in denying it a letter for its representation.

81. Sh. The mixed palatal *sh*, the cognate of *zh*, is clearly a single element. The blade of the tongue is well arched toward the hard palate, the teeth are nearly or quite closed, and the breath is thus expelled with much friction, giving a highly aspirated sound.

This sound is represented, in English orthography, by a great number of symbols, mostly digraphs, as *sh, ci, ti, ch,* etc.

82. Ch. The sound of *ch*, heard in *child, chin,* is the cognate of *j* and, like it, a compound difficult of analysis. The analysis, $ch=t+y$, is probably nearer the truth than the more common one, $ch=t+sh$; but its relation to either of these combinations is, doubtless, that of similarity rather than identity. In its formation the tip of the tongue is placed against the hard palate, and the teeth shut. The closed organs are then suddenly parted, and the escaping breath yields the sound of *ch*.

83. K. The sound of *k*, often represented by other letters, as *c, ch, gh,* is the only purely palatal aspirate in our language, though several are found in other languages, as the German.

For its production the soft palate is made to meet the base of the tongue, the nasal passages being also closed — the same position as that for its cognate *g*. When this complete closure is suddenly broken by the unvocalized breath, the sound of *k* results.

84. Th *(aspirate).* This sound differs from the subvocal *th* only in the lack of vocality. The tip of the tongue is placed under the edges of the upper teeth, and the breath is blown out between the teeth. It is a linguo-dental.

The substitution of this sound for that of *s*, constitutes the fault known as lisping. The simple direction for its cure is: Keep the tongue within the teeth while sounding *s*.
The remark with regard to the teaching of the sub-vocal *th* to foreigners, applies with equal force to this.

85. S. If the tip of the tongue be turned slightly upward near the upper teeth, as for *z*, and unvocalized breath be passed over it, the sound of *s* will be the result. It is a fine, sharp whistle. The common mouth, however, too often renders it as a coarse hiss.

This and its cognate, *z*, are sometimes called sibilants.

86. T. The letter *t* represents the sound of the puff of breath set free by the sudden parting of the middle closure of the mouth, that formed by the close contact of the tip of the tongue with the hard palate. If the sound of *d* be produced, and the breath be then blown out, it yields this sound, the pure lingual aspirate.

CHAPTER III.

PHONOTYPY.

87. Phonotypy is the art of representing speech-sounds to the eye by distinct and appropriate symbols.

This term, originally given to a particular system of speech-symbols, may now be appropriately applied to the whole art of phonetic representation.

88. The ancient Phœnicians are credited with making the first analysis of the sounds of speech, and with the adoption of a phonetic system of characters for the representation of the several sounds. This was an inconceivably great step in linguistic science, but one which has not been repeated. The present English alphabetic notation of sounds, or orthography, is no advance from the Phœnician system, but the reverse. It is, indeed, so imperfectly phonetic and so utterly unscientific as scarcely to deserve mention under this head — except for some consideration of its defects.

89. The defects of our English alphabet may be briefly specified.

1. For the representation of, say, forty-five sounds, it furnishes but twenty-six characters; and of these, three,

c, q, and *x,* are worthless, having no sounds of their own. Consequently, one letter, as *a,* must represent several sounds.

2. Our letters are unsteady in their powers, now representing one sound and now another, and often no sound at all. This is the source of great confusion.

3. Our orthography is inconsistent. Similar sounds find no similarity in their symbols, as *v* and *f,* for example. Single letters represent compound sounds, as *long i, long u,* and *j;* while digraphs represent single sounds, as *th, ph,* etc.

4. The letters do not represent the same sounds as in other languages. Thus our *long e* is represented in other languages by *i;* our *oo* sounds, uniformly by *u;* our *long a,* by *e,* and all in a far more symmetrical and consistent manner than in English.

In short, a more unsystematic, inconsistent, uneconomical method of representing speech than the present English orthography, is doubtless impossible to human ingenuity.

90. Various attempts have been made to devise a scientific and thoroughly phonetic system of sound-symbols. The most noted among these are the "Standard Alphabet" of Lepsius, Pitman's "Phonotypy," the "Palaeotype" and "Glossotype" of Alexander Ellis, and Bell's "Visible Speech." Many other phonotypic and phonographic systems have been employed by missionaries and short-hand reporters.

The Palaeotype of Ellis, presents a notation for about 400 distinct sounds; but perhaps the most ambitious of all these systems is the "Visible Speech" of Alex. Melville Bell, which undertakes to represent all possible human utterance by simple characters, picturing, as it were, the successive positions of the organs of articulation.

DIACRITICAL MARKS.

91. The inadequacy of the English alphabet is such that for the most ordinary purposes of orthoëpy, it has been found necessary to employ an auxiliary system of diacritical marks — guide-boards on the heads of our

bewildered letters — a needful makeshift to overcome the incapacity of our orthography for exact representation.

92. Webster's Dictionary employs the following marks:

Vowel marks . . .
- The Macron, ¯
- The Breve, ˘
- The Circumflex accent, ʌ
- The Tilde, or Wave, ∼
- Two Dots, ¨
- One Dot, ˙

Consonant marks.
- The Bar, —
- The Dotted Bar, ≐
- The Cedilla, ,

Most of these marks, and some others, are used in Worcester's Dictionary and in the Gazetteer and Biographical Dictionary, of Dr. Thomas, but not always with the same signification.

93. Significance of the diacritical marks. The MACRON and BREVE having been used from time immemorial to indicate the quantity of syllables, they are very naturally employed in all dictionaries to indicate the regular *long* and *short* sounds of the vowels.

The CIRCUMFLEX, or circumflex accent, long used to indicate "common" quantity, is employed by Webster to denote certain sounds of *a*, *e*, *o*, and *u*, before *r*—all long sounds. By Worcester the same mark is used to mark the *broad* sound of *a* and several substitute sounds, as *i* with the sound of *long e*, etc.

Two DOTS above the vowel are used by Webster to mark the *Italian a* only; by Worcester, for the same sound and also for the "short and obtuse" sounds of all other vowels when followed by a single *r* in accented syllables.

ONE DOT beneath a vowel is used by Worcester uniformly to indicate the obscure sounds of vowels in unaccented syllables, for which Webster, in general, employs no notation, depending upon the application of rules.

THE DOTTED BAR, used by Worcester to mark certain vowel sounds, is placed by Webster under *s* and *x* to indicate their use as sub-vocals (for *z* and *gz*).

For the signification of other marks, as the *tilde* or *wave*, the *cedilla*, etc., the dictionaries named may be consulted. A comparative table of the markings of the two, Webster and Worcester, would have been given but for the difficulty in obtaining proper type.

94. Spelling Reform. The grievous defects of our English orthography, as pointed out in Sec. 89, have become so evident and so burdensome as to enlist the most distinguished linguistic scholars of this country, and many in England, among the advocates of a reformed alphabet. The alphabet proposed by the Spelling Reform Association is thought to be the most feasible, as well as the latest, scheme yet proposed.

This alphabet, with a specimen of its use, is given below. A careful study of it is recommended.

It will be observed that modified forms of *a, o,* and *u,* are used for certain sounds of those letters, and diacritical marks are employed to denote long sounds where great accuracy is desirable. The consonant diagraphs in *h,* *(th, ch,* etc.) are retained. Duplicate characters for the sounds of *long e* and *a,* of *k, j, ng,* and *z,* are suggested for temporary use as "transition letters;" a transition stage being thought necessary before that of perfect phonetic representation.

ALPHABET OF THE SPELLING REFORM ASSOCIATION.

	Short.	Vauels.	Long.
I i,	it.	E ᵬ=ī, hᵬ, polīç.	
E c,	met.	Ꮐ a=c̄, potato, thêy, fare.	
A a,	at.	ā, fāre, (in America).	
Ꮐ a,	ask, (sᵬ Dictionaries).	ā, fär.	
Ө ө,	net, what.	ō, nēr, wall.	
O o,	wholly, (in New England).	ō, nō, hōly.	
U ʊ,	but.	ʊ, bʊrn.	
U u,	full.	ū, rūle, fool, mūv.	

Difthengs.—Ι ị=ai, fịnd, faind. ꭤU au, haus=house. ӨI ei, eil. Ꮜ ʮ er Ü ū=yu er iu, ūnit, yunit; mūᵬic, miuᵬic.

Surd.		Consonants.			Sonant.
P	p,	pet.	B	b,	bet.
T	t,	tep.	D	d,	did.
CH	ćh,	ćhurćh.	J, ɢ	j, g,	jet, ɡem.
C, K	c, k,q,	cake, cwit (quit).	G	g,	get.
F	f,	fit, filesofer.	V	v.	vat.
TH	th,	thin, pithy.	ᴅH	th,	Ɗhè, thè.
S, Ç	s, ç,	so, çent.	Z, ᴢ	z, a,	zone, iᴢ.
SH	śh,	śhè.	ZH	zh,	fûzhun.
WH	wh,	whićh, (in England).	W	w,	wè.

H   h,   hè.

L l, lo. Rr, rat. Yy, yè. Mm, mè.
N n, no. NG ng, Ŋŋ, kinɡ, iŋk.

Silabic:—l, nobl, nobla; m, spaɛm, spaamᴢ; n, tokn, toknᴢ.

  · Bi thè fonetic alfabet a ćhild ma bè tet thè art ev rèdinɡ, net fliłentli but wel, both in fonetic and in erdineri .buks, in thrè munths—ai, efn in twenti aura ev .thuro instrucśhun;—a task hwićh iᴢ rarli acomplirht in thrè yèra ev teil bi thè old alfabet. Hwet fathur er tèćhur wil net ɡladli hal and urnestli wurk fer this ɡrat bun tu edûcaśhun,—this pauurful maśhèn fer thè difûzhun ev neleɡ.

  An elturd erthegrafi wil bè unaveidabli efensiv tu thoᴢ hu ar furst celd upen tu ûᴢ it; but eni sensibl and censistent nû sistem wil rapidli win thè harti prefurenç ev thè mas ev riturᴢ.

  (The same, omitting the transition letters ᴢ, ɡ, ç, and the duplicate letter k, and using full forms for the diphthongs i and û).

  Bai thè fonetic alfabet a ćhaild ma bè tet thè art ev rèdinɡ, net fliucntli but wel, both in fonetic and in erdineri bucs, in thrè munths—ai, efn in twenti aurz ev thuro instrucśhun;—a tasc hwićh iz rarli acomplirht in thrè yèrz ev teil bai thè old alfabet. Hwet fathur er tèćhur wil net ɡladli hal and urnestli wurc fer this ɡrat bun tu ediucaśhun,—this pauurful maśhèn fer thè difiuzhun ev nelej.

  An elturd erthegrafi wil bè unaveidabli efcnsiv tu thoz hu ar furst celd upen tu yuz it; but eni sensibl and censistent niu sistem wil rapidli win thè harti prefurens ev thè mas ev raiturz.

CHAPTER IV.

ORTHOËPY.

95. Orthoëpy is the art of pronunciation, or the correct utterance of words. Its elements are Articulation, Syllabication and Accent.

ARTICULATION.

96. Articulation is that action of the tongue and other organs of speech by which each oral element receives its peculiar and proper character.

As the action of the organs is slight for vowel and great for consonant sounds, the chief labor of articulation is found in connection with the latter, some writers even limiting the term articulation to the execution of consonant sounds.

The word is derived from *articulus*, a little joint, and thus literally signifies *the jointing of speech*. The fitness of this term arises from the natural law of alternation in speech, the continual alternation of open and close sounds. This law may be illustrated by the following diagram:

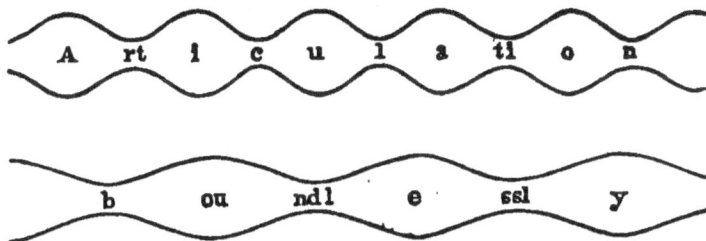

A rt i c u l a ti o n

b ou ndl e ssl y

Two or more consonant sounds may occur in the same joint; and two vocals capable of blending into a diphthong, may occur in the same node. Two consecutive vocals not thus blended, must be separated by a slight hiatus or pause.

97. Good articulation demands, in reading or speaking,—

1. The *exact* and proper utterance of each sound;

2. The utterance of *all* and only the required sounds;

3. The proper *separation* of the various sounds.

The corresponding errors in articulation are, 1. Bad enunciation; 2. Omission; 3. Blending.

SOME COMMON ERRORS IN ARTICULATION.

Analyze each of the following errors, and determine in what the error consists:

algebray	for	algebra.	*holler*	for	halloo.
Ameriky	"	America.	*hunderd*	"	hundred.
attackted	"	attacked.	*Id'no*	"	I don't know.
bile	"	boil.	*lickrish*	"	licorice.
bimeby	"	by and by.	*mushmelon*	"	musk-melon.
ketch	"	catch.	*miskeeter*	"	mosquito.
childern	"	children.	*mountanious*	"	mountainous.
drownded	"	drowned.	*nekked*	"	naked.
equil	"	equal.	*awnjiz*	"	oranges.
ellum	"	elm.	*pleg*	"	plague.
forrud	"	forward.	*piller*	"	pillow.
figger	"	figure.	*perty*	"	pretty.
Febyuary	"	February.	*pudd'n*	"	pudding.
f'rever n'ever	"	for ever and ever	*wich*	"	which.
git	"	get.	*yep*	"	yes.

98. The conditions of good articulation, and so of good pronunciation, are:

1. Flexibility and vigor of the organs of speech;

2. An exact knowledge of the peculiar character of each sound in the language;

3. A knowledge of the principles, or rules, according to which these sounds are combined; and,

4. Careful attention to the daily practical use of this knowledge, converting knowledge into skill.

Flexibility of the organs may be attained by suitable drill exercises, such as the utterance in rapid succession of the sounds *ah, ee, oo; it, ip, ik; hadē', hadī', hadō'*, with vigorous and exaggerated facial action.

Knowledge of the separate sounds may be acquired by the study of Chapter III. of this work; knowledge of their combination according to established rules or analogies, from the ensuing pages, or, more fully, from the standard dictionaries. Skill in application can be achieved only by careful and unremitting effort and attention.

THE ENGLISH SOUNDS.

99. Tables of the English sounds, as presented by Webster's Dictionary, are here given for convenience of reference. Each sound should be studied carefully with respect to its physical character, as set forth in the several sections of\Chapter III.

Chart of vocals and vowel substitutes.

Section.	Symbol	Name.	Key-word.	Substitute Symbol.	Key-word
51	a·	long	mate	e	prey
48	ă	short	mat		
50	â	circumflex	care	ê	where
40	ä	Italian	ah, far		
41	ȧ	short Italian	ask		
43	a̤	broad	awe	ô	nor
53	ē	long	me	ī	marine
49	ĕ	short	met		
59	ē̃	tilde	verse	ī	bird
54	ī	long	tine	y	my
52	ĭ	short	tin	ẏ	abyss
44	ō	long	bone		
42	ŏ	short	coffee	a	what
46	ōō	long	boot	{ Ǫ u·	do rude
45	ŏŏ	short	foot	{ ǫ ṵ	wolf push
55	ū	long	use, tune		
57	ŭ	short	up	ȯ	done
58	û	circumflex	urge		
47	ou		sound	ow	cow
56	oi		oil	oy	boy
	20				

100. Chart of consonant sounds.

(Cognates on the same line.)

Sec.	Symbol.	Key-word.	Substitute.	Symbol.	Key-word.	Substitute.	Sec.
	SUB-VOCALS.			**ASPIRATES.**			
69	b	bet		p	pet		79
76	d	dot		t	tin	ed, th	86
73	g	get		k	kit	c, ch, gh q	83
				h	hat		80
71	j	jet	ġ	ch	chin		82
64	l	lid					
68	m	mit					
65	n	not					
72	n	finger	ng				
63	r	rat, tar					
74	th	that		th	thin		84
67	v	vat		f	fat	ph, gh	78
61	w	woe		wh	when		77
62	y	yet					
75	z	buzz	s	s	sin	ç	85
70	z (h)	azure	si, zi	sh	shot	çh, c, ce, ci, si, ti, sch.	81

SYLLABICATION.

101. A syllable is a vowel sound which alone, or in combination with one or more consonant sounds, forms a word or a separable part of a word.

- The letter *l* is to be considered a vowel in the termination *ble* and sometimes in final *el*, the *e* being strictly silent. The letters *n* and *r* also sometimes perform the vowel office, as in *euchre, haven*, etc.

The longest syllable in the English language is the word *strength*.

102. Syllabication is the separating of a word into parts according to the number of its distinct vowel sounds.

Syllabication is the first step towards determining the pronunciation of an unfamiliar word. The difficulty of the process is much increased, in our language, by the frequency of silent letters and other irregularities.

The syllabication of words in spelling is of no value to the spelling itself, but it is of great importance, especially to children, *as an aid to pronunciation.*

103. Two general principles enter into syllabication, — the phonetic, or division with respect to smoothness and ease of utterance; and the etymological, or separation with respect to the derivation of the word.

Unfortunately for us no specific rules of much practical value can be given, so many exceptions arise from the conflict of the two principles named and from other causes.

SILENT LETTERS.

104. Silent letters, or those which are not direct representatives of sounds, constitute one of the chief hindrances to pronunciation. Many of these are as useless as they are annoying, while others perform somewhat the same office as diacritical marks, governing and indicating the sounds of other letters. Thus:

a. Silent *e* final usually indicates the long sound of the preceding vowel, as in *mete, fane.*

b. The doubling of a consonant usually indicates the short sound of the preceding vowel, as in *fallow, merry.*

c. A silent *u* after *g* indicates the hard sound of that letter as in *guide, vogue.*

In vowel digraphs, the silent letters serve to indicate the sound of the other, or active, vowels; though the great lack of consistency and uniformity in the influence which they exercise, renders them less useful to the learner.

105. Silent e occurs much more frequently than any other silent letter, and exercises a correspondingly great influence upon our orthoëpy and orthography. The following rules will be found of practical value:

RULE 1. *E* final is always silent except in monosyllables containing no other vowel, as *be, we,* and in classical or foreign words, as *Calliope, blasé,* etc.

RULE 2. *E* is usually silent in the termination *ed.*

Exceptions. (1) When preceded by *d* or *t*, the *e* is sounded from physiological necessity, as in *bounded, acted.*

(2) When *ed* is followed by *ly* or *ness*, the *e* has its regular short sound, as in *assuredly, blessedness.*

(3) A number of adjectives, mostly participial, have the short sound of the *e*, as *aged, beloved, blessed, crooked, cursed, dogged, hooked, learned, winged.*

As verbs or participles, however, they invariably drop the sound of the *e*.

RULE 3. *E* is usually silent in the termination *en*, as in *heaven*, which should be pronounced as nearly as possible in one syllable.

There are a few exceptional words, like *chicken, kitchen, hyphen;* and the *e* is sounded when preceded by *l, m, n* or *r*, as in *woolen, siren*, etc.

RULE 4. *E*, though usually sounded in the termination *el*, is silent in a few words, as *chattel, easel, hazel, ravel, shovel, weasel*, etc.

For full lists of the exceptions under the foregoing rules, see sections 57 to 61 of Webster's Unabridged and Academic dictionaries.

ACCENT.

106. Words of more than one syllable have one or more vowels pronounced with greater stress and clearness than the rest. This stress is called ACCENT. ·

The syllabication of a word being known, the next question presented is that of the location of accent.

The sounds of the letters occasion less difficulty. The syllabication and accent being known, the general rules, or analogies, of the language furnish guidance to the pronunciation of the great mass of English words, notwithstanding all that is said of the anomalous character of our language.

107. When two accents occur in the same word, they are of unequal force. The heavier one, in such cases, is called the PRIMARY ACCENT. The lighter is called the SECONDARY ACCENT. The secondary accent nearly always precedes the primary.

Nearly all words of more than four syllables have a secondary accent. Some very long words have two secondary accents, as *in-com' pre-hen' si-bil' i-ty;* but no accent ever falls beyond the sixth syllable.

A few of the simplest rules, only, are here given.

108. Rules for accent.

RULE 1. Simple words of two syllables, excepting *amen,* never have more than one accent.

> It is a very common error to pronounce such words as *combat, exile,* etc., with full stress on each syllable. This should be carefully avoided.
> A similar error consists in accenting two consecutive syllables in some words of more than two syllables, as in the words *exactly, idea,* etc., as sometimes heard.

RULE 2. In compound words each part retains its own accent, as in *morn'ing-glo'ry emp'ty-hand'ed.*

> When the component words of a compound are monosyllables, each retains its clear utterance, as when taken alone, but the greater stress is laid on that one which is descriptive or restrictive of the other, as in *seed'-corn, wheel'-horse.*
> When a compound has come into such common use, however, as to drop the hyphen, it is often accented like a simple word, as in *cup'board, high'land.*

RULE 3. Words which serve as verbs and also as nouns or adjectives, usually have the accent on the last syllable when verbs — in other cases, on the first syllable, as *contest',* verb; *con'test,* noun — *compound',* verb; *com'pound,* noun or adjective.

> Some words, however, as *address', express',* etc., do not change the accent to denote the part of speech. Many errors in pronunciation come from the failure to note these exceptions to the general rule.

RULE 4. All words ending in *sion* or *tion* have the accent on the syllable next to the last, the penultimate syllable, as in *presenta'tion.*

RULE 5. Words ending in *ical,* or *acal,* generally have the accent on the syllable next preceding, as in *ammoni'-acal, fin'ical.*

109. Monosyllables, when taken alone, or when at all emphatic, may be treated as if accented syllables. In common composition, however, monosyllabic pronouns, prepositions, conjunctions, and auxiliary verbs, and the articles, are usually quite unemphatic, and are then to be treated as unaccented syllables, receiving the same obscuration of the vowel sounds. See Secs. 117 and 118.

The article *the*, before a vowel sound, has the sound of *long e* so shortened as to resemble *short i*. Before a consonant sound, the sound of the *e* verges towards *short u*.

DRILL WORK. ANALYSIS AND APPLICATION.

110. Pronunciation is so greatly affected by habit that it becomes necessary, in the endeavor to eradicate ingrained errors and substitute correct for incorrect habit, to employ the most rigorous means for enforcing attention and assisting memory.

For this purpose, marking exercises, or drills in the application of diacritical marks; analysis of words by formula; and phonic spelling, will all be found useful.

Lists of words for such exercises are given at the end of the book. They are so selected as to serve a double purpose, all the words being such as are commonly mispronounced by the majority of speakers. These words should all be "looked up" in the dictionary, and the pupil not allowed to trust his past practice for anything.

The following is suggested as a suitable formula for the analysis of accented syllables:

1. **B-r-i-g-a-n-d** is a word of two syllables. The accented syllable is b-r-i-g. Its vowel sound is *ih* (Ĭ). The letter *i* is marked with a breve. The word is pronounced brĭg' and.
2. **C-a-u-c-a-s-i-a-n** is a word of three syllables. The accented syllable is c-a. Its vowel sound is *ae*. The letter *a* is marked with a macron; *si* has the sound of *sh*, and the word is pronounced cawcā' shun.

111. Phonic spelling. No course of instruction in orthoëpy can safely omit giving a considerable amount of drill in phonic spelling, or "spelling by sound." This exercise has especial value in the direction of articulation, tending to increase facility and accuracy therein.

For this purpose short, simple words should be used at first. The lists at the end of the book will furnish proper material for later work.

The teacher should insist upon the utmost exactness in the utterance of each successive sound, according to the descriptions of Chapter III., and upon proper syllabication.

RULES OF PRONUNCIATION.

112. Many of the rules or analogies which we unconsciously follow in every-day speech, are either so difficult of exact and at the same time simple statement, or so weakened by numerous exceptions, as to render their formal use difficult if not unprofitable. Accordingly it is thought best to present here only a very few of the simplest and most useful. No attempt is made to state all the exceptions existing.

113. Rules for consonants:

RULE 1. *C* when followed by *e*, *i*, or *y*, has the sound of *s*, as in *cede*, *city*.

The exceptions are *sceptic* (better spelled *skeptic*) and *scirrhus*, with their derivatives. In *sacrifice*, *sice*, *suffice*, *discern*, and their derivatives, *c* has the sound of *z*.
Ci and *ti*, before *ate* or *ation*, have the sound, of *shi*, as in *propitiate*, *pronunciation*.

RULE 2. *C* when followed by *a*, *o*, *u*, *l*, or *r*, and when it ends a syllable, has usually the sound of *k*, as in *cute*, *caustic*.

In *facade*, a French word, *c* has the sound of *s*.

RULE 3. *G* has its own or "hard" sound before *a*, *o*, *u*, *l*, or *r*, and at the end of a word, as in *gun*, *drug*.

The only exception is the obsolescent word *gaol*, and its derivatives.
G is also hard in the derivatives of words ending in *g*, as *druggist*, *craggy*. It has usually the sound of *j* before *e*, *i*, or *y*, but not always.

RULE 4. *N* has its second sound, known as *ng*, before the sounds of *k* and *g* hard, as in *finger*, *thankful*.

Exception: When the *g* or *k* sound begins an accented syllable, the preceding *n* has its common sound (*n* as in *no*), as in *concord'ance*, *tranquil'lity*.

4

RULE 5. *Q* has always the sound of *k*. It is always followed by *u*, which has the sound of *w*, as in *quart*, unless silent, as in *mosque, liquor*.

RULE 6. *X* has the sound of *gz* when followed by an accented vowel, as in *exact', exer'tion*. At the beginning of a word it has the sound of *z*, as in *Xerxes*.

A very common error in pronunciation consists in giving *x* the sound of *ks* before an accented vowel, in violation of the above rule.

RULE 7. *Y* has its own sound at the beginning of a syllable, as in *ye, beyond*. In other situations, and when it constitutes the syllable, it has the vowel office, as in *my, abyss, yclept*.

114. Vowels in monosyllables and accented syllables. In the statement of the following rules, monosyllables are considered as accented syllables.

RULE 1. An accented vowel at the close of a syllable has usually its long or name sound, as in *za'ny, pa'triot*.

RULE 2. An accented vowel followed by a single consonant (except *r*) in the same syllable, generally has its regular short sound, as in *man'ly, lin'en*.

RULE 3. An accented vowel in a syllable ending in silent *e* preceded by a single consonant (except *r*), has its regular long sound, as in *mice, debate*.

The three rules just given constitute the chief foundation of the "phonetic" method of teaching reading.

RULE 4. In accented syllables ending in *r* final or *r* followed by another consonant, and in derivatives of such words,

(1) *A* has its full *Italian* sound (ä), as in *barn, bar, debarred*.

(2) *E* has its third sound (ē), as in *fern, infer, inferred*.

(3) *I* has the sound of *tilde e*, (ĩ) as in *sir, stir, stirring*.

- (4) **O** has more commonly the sound of *broad a* (ô), as in *nor*, *storm;* but sometimes equals *circumflex u*, as in *word*, or *long o*, as in *ford*, *forge*.

(5) **U** has its third sound (ŭ) as in *cur*, *curt*, *incurred*.

(6) **Y** has the sound of *tilde e*, as in *myrtle*, *syrtic*.

RULE 5. An accented syllable ending in *r* doubled or *r* followed by a vowel, has the regular short sound of its vowel, as in *mirror*, *heroine*.

This rule is analogous to Rule 2.

RULE 6. In most monosyllables, and some other words, when followed by *ff*, *ft*, *ss*, *st*, *sk*, *sp*, and sometimes *nt* and *nce*, *a* has its *short Italian* sound, as in *pass*, *after*, *dance*.

RULE 7. **A**, when followed by *unch*, *und*, or *unt*, has its full *Italian* sound, as in *launch*, *laundry*, *haunt*.

This rule is not given because of its scope, but because it covers a class of words especially liable to abuse in pronunciation.

RULE 8. **U** preceded by *r* has the sound of *long oo*, as in *rule*, *ruin*, except in a few familiar monosyllables and their derivatives, as *run*, *rush*, which take the *short u*.

U never has its own long sound when preceded by *r*. This rule (Rule 8) is in fact an exception to Rule 1, but is worthy of separate statement.

115. Practice lists under the foregoing rules.
Let each word in the following lists be studied analytically, and referred to the proper rules in Sections 113 and 114. The use of the following formula, or some similar one, in recitation, is recommended:

(1) *C-h-a-r-a-c-t-e-r* is a word of three syllables. The accented syllable is *c-h-a-r*. It ends in *r* followed by a vowel; it therefore falls under Rule 5, Sec. 114, and the vowel *a* has its short sound (ă).

(2) *D-o-c-t-l-e* is a word of two syllables; *d-o-c* is the accented syllable. It ends in a single consonant, hence it falls under Rule 2, Sec. 114, and the vowel *o* has its short sound (ŏ). *C* is followed by *i*, and therefore has the sound of *s*, according to Rule 1, Sec. 113. The word is pronounced *dŏs'il*.

The syllabication and accent must first be determined
from the dictionary, if need be.

List 1.	List 2.
1. anemone	1. aunt
2. arable	2. craunch
3. agile	3. donkey
4. alternate	4. dauntless
5. Arabic	5. demoniacal
6. canine	6. erudite
7. caravan	7. fast
8. clangor	8. harass
9. currish	9. haunted
10. curry	10. jaunty
11. docile	11. miracle
12. enervate	12. narrow
13. extirpate	13. panegyric
14. horrid	14. paragon
15. larynx	15. ruthless
16. matron	16. rafter
17. myrmidon	17. saunter
18. orange	18. terrapin
19. peremptory	19. truculent
20. siren	20. tarry (verb)
21. sirup	21. tarry (adj.)
22. tartaric	22. taunt
23. tirade	23. ursine
24. virulent	24. wafted
25. whorl	25. zoology

VOWELS IN UNACCENTED SYLLABLES.

116. The vowels of unaccented syllables undergo,
in most cases, some obscuration or corruption of sound.
These changes take place, however, according to quite
uniform analogies, admitting of tolerably simple and
exact formulation.

The tabulation of these rules given below is adapted
from the discussion in Webster's Dictionary, by special
permission of the publishers.

It would be remiss not here to state the fact that, after all, the most marked difference between unrefined and refined speech, between boorishness and elegance of pronunciation, consists in the management of unaccented syllables. Here it is that vocalization and articulation are both liable to be defective, smothered, and bungling instead of clear, clean-cut, and ready. Increased elegance and effectiveness of speech will amply repay even protracted and painful self-discipline in this direction.

117. Unaccented syllables may best be separated into THREE CLASSES:

1. Those ending in a consonant.
2. Those ending in or consisting of a vowel (not silent *e*).
3. Those ending in *silent e* preceded by a consonant.

RULES FOR UNACCENTED VOWELS.

Class 1.

Vowels in Unaccented Syllables ending in a Consonant.

GENERAL RULE. The vowel has in strict theory its regular *short* sound, as in *entrust′, undo′.*

Caution. Carefully avoid the sound of *short u* in such words as si′lent, el′ement, etc.

Exception 1. A and o generally verge toward *short u,* as in big′ot, ramp′ant.

Exception 2. E, i, and y followed by r in the same syllable, have the sound of the second u in *sulphur,* as in read′er, ta′pir, sa′tyr.

Exception 3. Digraphs. Ai equals *short e* or *i,* as in mount′ain, maintain′; ei, ey and ie have the sound of *short i,* as in sur′feit, jour′ney; ow has the sound of *short u,* as in vig′orous.

Exception 4. Some Latin words have the *long* sound of the vowel in the terminal syllable, as in cri′ses.

Section 118.

	CLASS 2. *Unaccented syllables ending in a vowel (sounded).*	CLASS 3. *Unaccented syllables ending in silent e, preceded by a consonant.*
A.	1. Has usually its *short Italian* sound, as in *Cuba, America.* 2. When followed by another vowel, as in *aerial, chaotic,* has its *long* sound without the vanish. 3. In the terminations *ary* and *any*, sometimes verges toward *short e*, as in *literary.*	1. In verbs ending in *ate*, has its regular *long* sound, as in *delicate.* 2. In other cases, verges toward *short e*, as in *ultimate, preface.*
E.	Has its *long* sound slightly abridged, as in *event, benefit.* *Caution:* Avoid the sound of *short u* in all such syllables.	1. Equals *long e* slightly abridged, as in *obsolete, paraclete.* 2. In a few words, has its *short* sound, as in *college.*
I.	1. More commonly has its *short* sound, as in *direct, maniac.* 2. But in the initial syllables, *i, bi, chi, cli, cri, pri,* and *tri,* it has its *long* sound, as in *biology, criterion.*	1. Is more often *short*, but exceptions are so numerous that it is safest always to consult the dictionary. 2. Chemical terms in *ide* have the *i short*, as in *bromide.* 3. Names of minerals in *ite*, as in *steatite,* have the *i long.*
O.	Has its *long* sound without the vanish, as in *tobacco, option.* *Caution:* Avoid a secondary accent in the terminations *ory* and *ony.*	1. Has its regular *long* sound, as in *telescope.* 2. In a few words, has its *short* sound, as in *dialogue;* or that of *u,* as in *purpose.*
U.	1. Has its *long* sound slightly abridged, as in *accurate, usual.* 2. When preceded by *r,* equals *long oo,* as in *February, erudite* (not *eryoodite*).	1. Has its *long* sound slightly abridged, as in *gratitude, furniture.* *Caution:* Avoid the sound of *ch* or *j* (*creatyoor, not creacher*). 2. When preceded by *r,* equals *long oo,* as in *peruke,* but never when preceded by *t,* as in *institute.*
Y.	1. Usually equals *short i*, as in *vanity, hypocrisy.* 2. In verbs ending in *fy,* as in *magnify,* and in the words *multiply, occupy,* and *prophesy,* equals *long i.*	Usually equals *long i,* as in *anodyne.*

119. Drill-work. The analysis of syllables by formula will be found to afford not only an effective means of enforcing application and retention of the forgoing rules, but also a *logical drill* scarcely surpassed by any other in the whole round of school work.

The following formula, or any similar one, may be used in connection with the practice lists here given:

FORMULA.

(1) In the word *c-o-m-b-a-t*, *b-a-t* is an unaccented syllable ending in a consonant. It falls under class 1, exception 1: "*A* and *o* usually verge toward *short u.*" The syllable is pronounced ———.

(2) In the word *p-a-r-a-c-l-e-t-e*, the second *a* is an unaccented syllable consisting of a vowel. It therefore falls under class 2: "*A* usually has its *short Italian* sound." The syllable is pronounced ———.

C-l-e-t-e is an unaccented syllable, ending in silent *e* preceded by a consonant. It therefore falls under class 3: "*E* has its *long* sound slightly abridged." The syllable is pronounced ———.

PRACTICE LISTS. In the following words, analyze the accented syllables according to formula in Section 115; the unaccented, according to that just given. The words are arranged in a progressive order corresponding to the order in which the rules are presented.

LIST 1.	LIST 2.
1. nomad	1. estimate
2. solemn	2. communicative
3. government	3. chimera
4. character	4. direction
5. combatant	5. irascibility
6. maintain	6. hospitality
7. silent	7. chloride
8. sleeplessness	8. civilization
9. indifferent	9. telephone
10. cathedral	10. respiratory
11. banana	11. polonaise
12. aorta	12. gondola
13. separate	13. orthoepy
14. comrade	14. obligatory

LIST 1.

15. secondary
16. incomparable
17. elementary
18. paraclete
19. benefited
20. remunerative

LIST 2.

15. undisputably
16. garrulous
17. literature
18. virulently
19. inopportunely
20. phonotype

120. Words commonly mispronounced.

It is urged that the student make himself accurately acquainted
with the pronunciation of all the words in the following lists.
They will also furnish material for drill work in connection with
preceding sections of the work.

I.	II.
1. abdomen	1. Calliope
2. acclimate	2. Canaan
3. acoustics	3. carbine
4. address	4. Caucasian
5. Adonis	5. chastisement
6. albumen	6. coadjutor
7. allies	7. combatant
8. allopathy	8. comparable
9. allopathic	9. construe
10. almond	10. creek.
11. alternate	11. cupola
12. apparatus	12. cushion
13. area	13. deficit
14. aroma	14. depot
15. aspirant	15. discourse
16. banana	16. dishonest
17. behemoth	17. docile
18. benzine	18. donkey
19. blatant	19. envelope
20. bombshell	20. enervate
21. bouquet	21. erring
22. bonnet	22. errand
23. brigand	23. exemplary
24. bronchitis	24. exquisite
25. brooch	25. extol

III.

1. facade
2. February
3. finále
4. finance
5. forgery
6. frontier
7. franchise
8. fugue
9. gape
10. gauntlet
11. giraffe
12. glámour
13. gladíolus
14. granary
15. homœopathy·
16. hydropathy
17. indisputable
18. inquiry
19. integral
20. isolate
21. ĩsotherm
22. italic
23. jaguar
24. jaundice
25. jugular

IV.

1. lath
2. lamentable
3. leisure
4. lïen
5. lycéum
6. machïnãtion
7. mãnïacal
8. multiplicãnd
9. näïvetę'
10. national
11. nŏm'ad
12. obligatory
13. Oríon
14. õrotund
15. Palestine
16. parent
17. patriotism
18. pátron
19. peremptory
20. photographer.
21. placard
22. portent
23. porcelain
24. precedence
25. precēdent

V.

1. prelate
2. presentation·
3. produce (noun)
4. pronunciation
5. pyrámĭdal
6. raillery
7. rapine
8. raspberry
9. rational
10. recess
11. recĭtatĭve
12. recruit

VI.

1. squalĭd
2. squalor
3. stalwart
4. tále
5. täunt
6. telegrapher
7. Thalia
8. tĭny
9. tomạto
10. trãŋquil
11. tribúne
12. trụculent

V.

13. rĕpárablẹ
14. research
15. resōurce
16. respiŕatory
17. rĭbáld
18. roͅmánce
19. rooͅt
20. săc̄rifice
21. sagā́c̄iouͅs
22. sälve
23. scarcėly
24. sēine
25. sha'n't

VI.

13. trўst
14. vagāry
15. vălet
16. vĭeͅar
17. wáter
18. wōn't
19. wräth
20. yacht
21. yĕarling
22. yōlk
23. you
24. zōdiͅacal
25. zoölogy

INDEX.

A.

	Page.
A, as in care	26
A, Italian sound of	23
A, long sound of	26
A, short sound of	25
A, short Italian sound of	23, 51
A, broad sound of	24
A, in unaccented syllables	53, 54
Accent	46, 47
Alphabet, English	36, 37
" " defects of	36, 37
" Phœnician	36
" universal	37
Analysis of syllables	48, 51, 55
Articulation, process of	14
Articulation, definition, demands, etc.	41, 42
Aspirates	18, 34, 44
Atonics	18

B.

B, sound of	31
Breathing, process of	11
Breve	38
Broad a	24
Bronchi	8

C.

C, rules for sound of	49
Ch, sound of	35
Chords, vocal	10
Circumflex	38
Circumflex a	26
Circumflex u	28
Classification of oral elements	17
" " "	19
Classification of unaccented syllables	53
Cognates	18
Common errors in articulation	42

	Page.
Consonants	18
Consonant sounds, table of	44
Correlative sounds	21

D.

D, sound of	33
Defects of English alphabet,	36, 37
Descriptive phonology	23, 36
Diacritical marks	37, 38
Diagram of oral elements	22
Diagram of articulation	41
Diaphragm	6, 7
Digraphs	19
Diphthongs	19, 27
Dots, diacritic	38
Dotted bar	39
Drill work	48, 51, 55

E.

E, long sound of	27
E, short sound of	25
E, tilde	28
E, as in there	26
Ear, the	15
Elementary sounds	17, 43
English sounds	43, 44
Epiglottis	11
Errors in articulation	42
Ew, as in new	27
Expiration	12

F.

F, sound of	34
Formula for accented syllables	48, 51
Formula for unaccented syllables	55

G.

G, sound of...............33, 49
Glottis....................... 10

H.

H, sound of 34

I.

I, long sound of............ 27
I, short sound of........... 26
I as in *pique* 27
I as in *girl*................. 28
Inspiration 11
Intercostal muscles......... 7
Italian *a* 23

J.

J, sound of................. 32

K.

K, sound of................. 35

L.

L, sound of................. 30
Labials 19, 22, 31
Larynx 8,9
Linguals......... 20, 22, 28,33
Lips.................... 14, 20
Lists of words 52, 55, 56,58
Long vowel-sounds...... 20,21
Long *a* 26
" *e* 27
" *i* 27
" *o* 24
" *oo* 24
" *u*................... 27
Lungs 6

M.

M, sound of............... 31
Macron 38
Mixed diphthongs 27
Monosyllables 47, 48

N.

N, nasal sound of.......... 30
N, palatal sound of...... 32, 49
Nasal passages 13, 20
Ng, sound of.............. 32
Noise....................... 16

O.

O, long sound of......,.... 24
O, short sound of 23
O as in *word* 28
O as in *son*................. 28
Obstruction of sound 31
Oi, Oy, sound of.......... 28
Oral elements 17, 22
Organs of speech 13
" " voice 6
Orthoëpy, definition of ... 5, 41
Orthography, " 6
Orthography, English, de-
fects of...............37, 38
Ou, Ow, sound of.......... 25

P.

P, sound of................. 34
Palate................ 13
Palaeotype. 37
Palatals 19, 22, 25, 32
Pharynx 11
Phœnician alphabet........ 36
Phonetics 5, 16
Phonetic print.............. 40
Phonic spelling............. 48
Phonology, definitions.... 5, 16
" descriptive .. 23, 36
Phonotypy................ 5, 36
Practice lists.. 51, 52, 55, 56–58
Pronunciation, rules of.. 49, 50

Q.

Q, sound of................. 50
Quality 21
Quantity................... 20

R.

R, effect of on vowel-sounds,
50, 51
R, sounds of 29, 30
Respiration 11
Rules for accent............ 47
" accented vowels, 50, 51
" unaccented vowels,
53, 54
" consonants........ 49
" pronunciation.. 49, 50
" silent *e* 45, 46

S.

S, sound of................ 36
Semi-vowels.........21, 29, 30
Sh, sound of.............. 35
Short vowel-sounds......20, 21
   " a 25
   " e................... 25
   " i................... 26
   " o................... 23
   " oo.................. 24
   " u 28
Silent e................45, 46
   " letters, uses of....... 45
Sound 16
Speech 17
Spelling by sound.......... 48
Spelling reform.........39, 40
Sub-tonics 18
Sub-vocals 18
Syllabication 44, 45
Syllable, definition......... 44

T.

T, sound of................ 36
Table of vocals............ 43
   " consonants....... 44
Table of rules for unaccented
   vowels.... 54
Teeth ...,............14, 20
Terminations in ed, el, etc 45, 46
Th, aspirate............... 35
Th, sub-vocal.............. 33
The, pronunciation of..... 48
Tilde, e................... 28
Tone 16
Tonics 18
Tongue 13, 14
Trachea 8
Trigraphs 19

U.

U, long sound of.......... 27
U, short sound of.......... 28
U as in urge.............. 28
U preceded by r, rule....... 51

Unaccented syllables, rules 53, 54
Universal alphabets... 37
Uses of silent letters....... 45
Uvula..................... 13

V.

V, sound of................ 31
Visible speech............. 37
Vocal chords 10
Vocalization 12
Vocal organs.............. 6
   " physiology 5, 6
Vocals, or vowel-sounds, 17, 20
           43
Voice..................... 16
Vowel 18
Vowel sounds 17, 20, 23, 43
   " " rules for....50, 51
Vowel in unaccented sylla-
   bles...................52–54
Vowel substitutes.......... 43

W.

W, sound of............... 29
Webster's markings........ 38
Wh, sound of.............. 34
Words commonly mispro-
   nounced 56–58

X.

X, sound of 50

Y

Y, as a consonant.......29, 50
Y as in abyss............. 26
Y " my................ 27

Z.

Z, sound of................ 33
Zh, " 32

ELEMENTS OF ENGLISH ANALYSIS.

ILLUSTRATED BY A NEW SYSTEM OF DIAGRAMS.

By STEPHEN H. CARPENTER, Prof. of English in the University of Wisconsin.

This book, the result of the author's experience in the class room, is designed to assist students, by a System of Diagrams, in obtaining the outline structure of sentences, which a thorough knowledge of English grammar demands, thus fixing in the eye and mind the principles of analysis, a correct knowledge of which, as a rule, is wanting among students.

Price, in boards, 25 cents. Mailed on receipt of price.

RECOMMENDATIONS:

"I am impressed that you have made grammatical analysis so plain that the learner will find the work an aid to mastering rather than a hindrance to ordinary understanding of the language, as many systems of analysis are hindrances. Your book presents the first diagrams which I have seen that are not obstacles to my understanding. I trust you will take special means to bring this book to the attention of teachers."—PRESIDENT PARKER, *of the River Fall Normal School*, (Wis.)

"Mrs. Bateman adopted its suggestions at once, and finds it a decided improvement upon former schemes. * * * We shall adopt it for use in our next class."—PRESIDENT ALBEE, *of the Oshkosh Normal School*, (Wis.).

"Most of our works on analysis are too complicated and prolix; you have happily avoided that error. The statements are clear and concise; the definitions good; and the illustrations excellent. The diagrams are, in my judgment, a great improvement upon those in most works of the kind. Chapter xiii (on the infinitive) clears up a difficulty met by every teacher of grammar. You are to be congratulated upon your success in the production of a work so excellent, so timely, so comprehensive, in so small a compass, and that will meet a want so generally felt."—PROF. B. M. REYNOLDS, *of the La Crosse High School*, (Wis.).

OPINIONS OF THE PRESS:

"This is the title of a little book on purely sentential analysis, which appears to us worthy of high praise for its clearness, methodical arrangement, accuracy and *brevity*. The simple yet helpful diagrams—the chief original feature of the work—admirably illustrate the golden mean in this kind of ocular aid, and can give no offense to the most fastidious grammarian. The nomenclature is that in common use, the examples for practice are well selected and sufficiently numerous, and in paper and typography the book is faultless. We heartily commend it to teachers and school officers."—*Wis. Journal of Education,*

"The design of this little work is to explain the construction of the English sentence upon philosophical principles, and at the same time to exhibit this structure to the eye by a system of diagrams that will present the anatomy of the sentence with no confusing details. This little work, in few pages, seems to cover the whole ground, and must be commended at least for its brevity. It is the work of one who has encountered and surmounted the difficult task of bringing syntactial analysis within the comprehension of any ordinarily bright English scholar. It is the result of years of practice in the class room—of that kind of practice that 'makes perfect.'" The mechanical execution of the book is in keeping with its intrinsic merits, being printed on new, clear, and beautiful-faced type, while the diagrams were cast expressly for this work.—*Wisconsin State Journal.*

W. J. PARK & Co., Publishers,
Madison, Wis.

A System of Punctuation:

FOR THE USE OF SCHOOLS.

By C. W. BUTTERFIELD.

PUBLISHED BY WM. J. PARK & CO., MADISON, WIS.

This is a concise treatise on Grammatical and Rhetorical Punctuation, intended especially for the use of schools. Notwithstanding this, it is also adapted to the requirements of professional and business men who desire to write or correspond without fear of misapprehension or mistake.

Extracts from the Preface.

"That the principles of Punctuation are subtle, and that an exact logical training is requisite for the application of them is claimed by some writers; others, however, think the subject is founded largely in caprice,—that its rules are, to a great extent, conventional. Neither of these views is the correct one. The laws governing the uses of the various characters of Punctuation, require, it is true, a continued exercise of judgment in their reduction to practice; for any principle, however plain, necessitates some action of the mind to fully comprehend it. But these laws have, in a large degree, become fixed by established usage.

· · · "But little effort is requisite to convince pupils of the importance of Punctuation; and it is only necessary to bring the subject before them in a systematic manner, to enlist them at once in the study. It has been the object of the author so to treat it as to lead the student, step by step, and with comparative ease, from its simplest, to a just comprehension of its most difficult principles. Accuracy in definition, clearness in arrangement, and perspicuity in language, have been attempted in the presentation of the various rules; with what success is left to the judgment of the public."

The work has two principal divisions. The first treats of the nature and uses of the various characters of Punctuation. The second contains promiscuous examples for their application. What has been learned of the *theory* of Punctuation, can thus be put in *practice*.

RECOMMENDATIONS.

Butterfield's System of Punctuation is at once concise and full. It is fitted to do a good work.—DR. JOHN BASCOM, *President of the University of Wisconsin.*

I have examined Butterfield's Punctuation with a great deal of interest. I have long felt the need of something of the kind. I shall gladly call the attention of teachers to it.—KENNEDY SCOTT, *County Sup't of Columbia Co., Wis.*

Butterfield's Punctuation is clear and concise in statement, methodical in treatment, and sufficiently comprehensive in scope to meet the wants of the general public. Its general use in our schools would be of great benefit.—DR. S. H. CARPENTER, *University of Wisconsin.*

Butterfield's System of Punctuation should be in the hands of every teacher and student.—PROF. R. B. ANDERSON, *University of Wisconsin.*

I have examined Butterfield's Punctuation and am pleased with it. I shall recommend it to teachers.—H. RICHMOND, *County Sup't of Green Co., Wisconsin.*

ANALYSIS AND EXPOSITION

OF THE

CONSTITUTION OF WISCONSIN,

BY A. O. WRIGHT.

Published by DAVID ATWOOD, Madison, Wis.

We have just issued the sixth edition, revised and corrected, of this valuable little work — By A. O. WRIGHT. The press of the State have noticed this book in a favorable manner. It cannot fail to become very useful to the people, and as soon as known, will occupy a place in the library of every citizen of the State, who desires to understand thoroughly our Constitution.

The following letters have been received from gentlemen well known as distinguished educationists in this State:

[*From* Prof. SAMUEL FALLOWS.]

I have read, with care and interest, the book on "The Analysis and Exposition of the Constitution of Wisconsin," by Mr. A. O. WRIGHT.

It seems to me to be admirably adapted to the needs of teachers and scholars in the common schools of Wisconsin, and a valuable work of reference for all who wish a clear and succinct treatise on our State Constitution.

The analyses are thoroughly made — the language is plain and simple — the citation of authorities ample and correct.

If any inaccuracies should be discovered in this first edition, I believe they will be found to be of a minor nature. They can easily be corrected in a subsequent edition.

Mr. WRIGHT deserves great praise for the excellent manner in which he has executed a difficult task. I think his work will have a speedy, wide and permanent circulation in the state. SAMUEL FALLOWS, Superintendent of Public Instruction.

[*From* O. M. CONOVER, *Supreme Court Reporter.*]

The provision of our present school laws which requires the Constitution of the United States and that of our own State to be taught in our public schools, seems to me of great importance. It is clear, too, that the object can not be well accomplished without the publication of editions of those instruments, accompanied by simple, yet correct expositions of their language and scope.

I have read with some care a portion of the little work of Mr. WRIGHT upon the Constitution of Wisconsin. It appears to me to have been prepared with a very correct appreciation of the wants of our public schools, and to be in general, correct in its exposition of the spirit and meaning of the constitution, and well adapted to excite an interest in the study of that instrument, and to impart useful information in regard to it. Doubtless, a severely critical examination of the work will lead to the detection of some errors or deficiencies, which may be corrected in future editions. This is the common fate of first editions of school books which pioneer the way in any new department of instruction. But the general aim and plan of the volume are so good, and its execution is marked by so much intelligence and care, that it can hardly fail, I should think, to come into general use in the schools of Wisconsin. O. M. CONOVER, Supreme Court Reporter.

I think the work worthy of being in every school in the State, and in the hands of every voter. — J. T. LUNN, *Co. Sup't Sauk Co.*

www.ingramcontent.com/pod-product-compliance
Lightning Source LLC
Chambersburg PA
CBHW021524090426

42739CB00007B/765